On Quantum Mechanics

Greg Feild

October 15, 2019

a greg feild theory   :)

↺ ↻

gf

About the author:

    I earned a PhD in experimental high energy physics from the Pennsylvania State University working on HERA at DESY in Hamburg, Germany studying photoproduction and deep inelastic scattering in electron-proton collisions.

    I did my postdoctoral studies with Yale University working at Fermilab on the CDF experiment at the Tevatron. My primary research interest was particle hadronization in quarkonium production in proton-antiproton collisions.

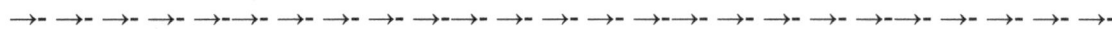

Despite their good intentions, those very people who believe themselves to be the most faithful spokesmen for their predecessors transform the thoughts which they want simply to repeat; methods are modified because they are applied to new objects. If this movement on the part of philosophy no longer exists, one of two things is true: either the philosophy is dead or it is going through a "crisis." In the first case there is no question of revising, but of razing a rotten building; in the second case the "philosophical crisis" is the particular expression of a social crisis, and its immobility is conditioned by the contradictions which split the society. A so-called "revision" performed by "experts," would be, therefore, only an idealist mystification without real significance. It is the very movement of History, the struggle of men on all planes and on all levels of human activity, which will set free captive thought and permit it to obtain its full development.

    Those intellectuals who come after the great flowering and who undertake to set the systems in order or use the new methods to conquer territory not yet fully explored, those who provide practical applications for the theory and employ it as a tool to destroy and to construct --- they should not be called philosophers. They cultivate the domain, they take an inventory, they erect certain structures there, they may even bring about certain internal changes; but they still get their nourishment from the living thought of the great dead. They are borne along by the crowd on the march, and it is the crowd which constitutes their cultural milieu and their future, which determines the field of their investigations, and even of their "creation."

These *relative* men I propose to call "ideologists."

Jean-Paul Sartre
Search for a Method

Abstract:

In this book we examine quantum mechanics in light of
The Universal Model of Our Sinister Universe.

→- →- →- →- →-→- →- →- →- →-→- →- →- →- →-→- →- →- →- →-→- →- →- →- →-

    a thumbnail sketch
    a jeweler's stone

                                        a mean idea to call my own

    →- →- →- →- →-

    standing on the
    shoulders of Giants

    leaves me cold

    a mean idea
    to call my own

    →- →- →- →- →-

    everybody hit the ground
    everybody hit the ground
    everybody hit the ground
    everybody hit the ground

                -- R.E.M.
                King of Birds

                                                      Go Dawgs !

Prelude:

In our last book, "On Math, Physics and Metaphysics", we had to admit to a bit of inadvertent plagiarism. I knew I was quoting someone almost verbatim, but I could not remember who it was or where I had absorbed this brilliant piece of prose!

Since we are copping to inadvertent plagiarism, a similar thing happened in the book "On angular momentum, acceleration, and absolute motion" (not the best book ever!).

In the section "Force and acceleration", in the paragraph beginning

"A model where everyone is at rest ...",

the words are almost certainly not mine. While I was typing it in, I felt like I had read it somewhere before, but it sounded like me! I apologize profusely to this cunning linguist. I will probably remember the source one day. In the meantime, perhaps this person may come forward to claim their work.

As far as copyright concerns, and our many book excerpts; we can only argue 'fair use', and their employment in, what we hope is, academic and social advancement.

I understand we've been playing it a little fast and lose here!

But, we are a just a teeny, tiny, little, itty-bitty, Mom and Pop organization. : (

On the bright side, maybe these folks will get the universal model 'bump'.

:)

let's not be litigious !

The phrases "burr in the ass of progress" and "fart crosswise" were coined by the brilliant wordsmith and high school shop teacher Curtis H. Shore, Philipsburg, PA.

Preface:

In our last book, we felt it necessary to say some unkind things, make some unpleasant observations, and resort to coarser language than we'd prefer.   But . .

People either value Knowledge (knowledge for its own sake), or they do not.

People either take "human advancement" seriously, or they do not.

People either define human advancement as the freeing ourselves of superstition, in-group social conditioning, knee-jerk behavior, and tribal inclinations, or they do not.

Some things need to be said.

Unfortunately, in our brave, new, pathetic age of ignorance, people only respond to hysterical screaming and hyperbole.

Why are people constantly shouting and screaming at each other, 24/7, on the television?

That's the way people behave in real life !

:(

    turn off the tv. read a book

    read a 'real' book

    Let the healing resume !

:)

Quantum Physics: 3.7 Operators, Eigenvalues, and Measurements

The state function Ψ contains all the information that can be known about the system. The behavior of Ψ is determined by the Schrodinger equation,

$$H\Psi = i\hbar \,\partial\Psi/\partial t$$

Each measurable dynaminical quantity is represented by a Hermitian operator Ω with a complete set of orthonormal eigenfunctions Ψ_n

$$\Omega\Psi = \omega_n \Psi_n$$

Since the Ψ_n are complete, the state function can be represented by the expansion

$$\Psi = \sum_n C_n \Psi_n, \quad C_n = \int \Psi_n^* \Psi$$

The possible results of a measurement of Ω are the eigenvalues, and the probability of obtaining ω_n is $|C_n|^2$.

-- Rolf G. Winter

Nothing spooky so far !

The expansion of states above, represents a mathematical superposition of states, and *not* a physical superposition of states. The particle or system is always in a well defined, explicit physical state.

The statistics of quantum mechanics behave exactly like those of any invisible macroscopic physical system, and/or any large collection of correlated variables.

Physicists do not understand statistics ... Who does ?

If you are reading a physics paper and encounter the word 'Bayesian', commit it to the flames!

If you happen across the phrase 'neural network' in a physics book, well,

a fool and their money are soon parted

Introduction:

The world is deterministic. Elementary particle interactions are deterministic.
Physics is deterministic. Quantum mechanics is deterministic.

However, the fact that observer A is now making a measurement at time T,
is not predetermined, although the outcome of their measurement may well be!

So much for determinism.

Laplace's claim to be able to calculate the course of the universe would be undoubtedly true, and assured, were it not for the existence of conscious agents: human beings, dogs, two thirds of the Animal Kingdom, and yes, even the celebrity physicist!

People screw everything up.   They always go *too far*.

When ideas become isms; that is the death of thought.

When abstract ideas become *undiscoverable*, yet certain, concrete realities;
thinking is dead; not only dead . . . but D.O.A.   Is such thought preordained ?

Most of history's most competent philosophers have indicated, or concluded, that it is nigh impossible to truly define thinking (What is it? How do you do it? How do you distinguish thinking from imagining, feeling, or doing syllogisms?), and some would say we are incapable of rational thinking or saying anything meaningful at all!   (They obviously went "too far" !)

What do *you* think ?

Follow yourself !

The new tao of physics:

The seven stages of grief, and

The twelve step recovery program.

Quantization:

   Elementary particles are fundamental units of angular momentum 'oscillating' in two (the photon) or three (the leptons) dimensions, at frequencies determined by the particle energies and momenta; or, more precisely, vice versa !

   The fundamental unit of angular momentum is Planck's quantum of action, h.

   The oscillations of the quantum of action are manifest as the sinusoidal projection of the particle angular momentum along the direction of travel.

   This is the origin, or nature, of the quantization of energy and linear momentum, and the reason for the uncertainty principle.

   A particle's position is only available for measurement, or 'real' interaction, for about half the time of any observation period.

   The other half of the time, the particle would rather be bouncing off the walls!

   These ideas have been explored several times before in previous books, and as we are as loathe to write any more about it (i.e. cut and paste !) as the audience is, most likely, to read it --

Please see "Revenge of the Sinister Universe" and "On Math, Physics, and Metaphysics" for the matchbook, and back of the envelope, summaries.

   The only properties of the the elementary particles are spin, parity (!), mass, charge, and helicity. These properties completely define the particle and all particle interactions. These quantities are all measurable and all have physical, if not identical, counterparts in the 'macroscopic' world.

   There are no quantum labels. No flavors, no colors, no weak charge, no gluons, and no eternally confined partons or quarks. Partons are electrons and neutrinos (the only new thing under the sun . . .). The neutrino has been hiding inside the neutron the whole time !

   Particles cannot be in two places at once, they do not become entangled, and they do not exist in a superposition of states, despite 'measurements' to the contrary.

garbage in, garbage out

(still healing)

Reverse pilot wave theory:

For a free particle, $x_0 = 0$, $v_0 = v$

$x = v_0 t \sin(px - Et)/\hbar$ $\rightarrow$ $x = A \sin(px - Et)/\hbar$

$p = mv_0 \cos(px - Et)/\hbar$ $\rightarrow$ $\dot{x} = A(E/\hbar)\cos(px - Et)/\hbar$

The variables x and p are out of phase by $\pi/2$.

In analogy with the case of the classical, one dimensional, simple harmonic oscillator, we can display the variation of the position and the velocity (x, $\dot{x}$) of the particle as a phase space diagram. This phase space diagram will show the relative variations in the ability to measure the position or the momentum of the particle at any time t, and/or the amount of linear momentum available for interaction at any point x, at any time t.

Time, again! Particles move, and systems evolve, in time, and *over* time.
Why doesn't time run backwards?   **_Time_** doesn't run *any which way.*
Crap in motion, stays in motion.   Anyroad ...

For a free particle, we can write

$x^2/A^2 + (\dot{x})^2/A^2(E/\hbar)^2 = 1$; $A = 1/\sqrt{2}$ (?) $\Rightarrow 1$ (why not ?)

$x^2 + (\dot{x})^2/(E/\hbar)^2 = 1$

$d(\dot{x})/dx = -E/\hbar \cdot x/\dot{x}$

For a free particle, the phase space path will consist of closed circles in the x,$\dot{x}$ plane.

For a free particle, "we also know" (17), $T = I\omega^2$

Lagrange's equation is

$\partial T/\partial \theta - d/dt(\partial T/\partial(\dot{\theta})) = 0$; $\dot{\theta} = \omega$; $\omega = E/\hbar$

$\rightarrow$ $d/dt(2I\omega) = 0 \rightarrow I\, d\omega/dt = 0$

Which is Newton's second law. More generally

$\tau = dI/dt \cdot \omega + I\, d\omega/dt$

The hydrogen atom:

Electron orbital decay in the hydrogen occurs via the familiar process of *bremsstrahlung* as illustrated in Figure 1.

Everything is *bremsstrahlung*. Gravitational waves are *bremsstrahlung*.

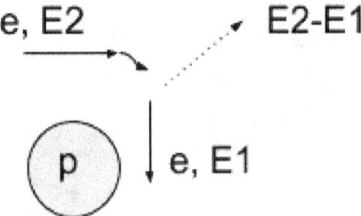

Figure 1: Photon emission in hydrogen electron orbital decay depicted as braking radiation.

An electron in an excited orbital can be considered to be moving tangential to said orbit, and very fast too! The electron is deflected from its path due to the influence of the 'infinitely' heavy proton. (this is the universal complementary principle - sometimes, the proton moves, sometimes, not!)

The emitted photon is represented by the dotted arrow, $E = E_2 - E_1 = h\nu$.

The photon has spin 1, so, of course, for the hydrogen atom as a whole, $\Delta l = 1$.

The electron has shed kinetic energy and is closer to (and as close as it can get!) to its rest mass.

The electron orbits are quantized because the electron *must* form closed elliptical paths.

This is the universal model of electron orbital decay.

The mirror image of our model of beta decay!

Beta decay:

In our model, the proton is a bound state of two positrons and an electron, and the neutron is the bound state of an electron, a positron, and an electron antineutrino.

These assignments would indicate that the proton is 'antimatter' and the neutron 'matter', and thus they would have opposite parity. In our model, the electron, proton, and neutron cannot all have parity = +1 . . .

The strong coupling constant is (6)

$$\alpha_s = \alpha^0_s (1 + \tfrac{1}{2} v^2/c^2 + \tfrac{3}{8} v^4/c^4 + \ldots) \tag{1}$$

$$\alpha^0_s = (G/4\pi\varepsilon)^{1/2} (2m_e e/\hbar c) \tag{2}$$

The weak coupling constant is

$$\alpha_w = (m_\nu^2 G)/(\hbar c)\, (1 + (v/c)^2 + (v/c)^4 + \ldots) \tag{3}$$

Once again (5), we will try to construct a strong force potential using Hooke's law. The force between two charged leptons (i.e. electrons and/or positrons) inside the nucleon is

$$F_{1,2} = \pm\, \alpha_s / R^2 \tag{4}$$

If we assume R is constant, and neglect the constant term in equation (1), then for an electron and a positron, we have

$$F = \tfrac{1}{2}\, \alpha^0_s v^2/c^2 \tag{5}$$

$$U = -\, \alpha^0_s v/c^2 \tag{6}$$

Next, we will want to make the usual expansion of v about the equilibrium velocity v_0, and cue the magic, eventually we can arrive at a covariant formulation in terms of

$$\Delta V^2 = (v - v_0)^2.$$

The proton will be a complicated interaction of two (three ?) strongly coupled harmonic oscillators, which is why we choose to focus on the neutron instead; in particular, we want to investigate the nature of beta decay.

The universal model of beta decay is depicted schematically in Figure 2.

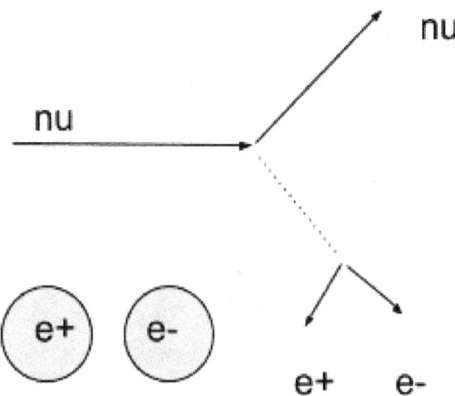

Figure 2: An antineutrino is in a loosely bound orbit around a relativistic positronium nucleus. The nucleus is bound by the potential of equation (6). The nucleus itself is unstable, so the whole thing is a house of cards!

A neutrino, loosely bound 'inside' the neutron, radiates a ~real photon, as if it were free, with the nucleus providing the recoil, of course. The photon decays, or splits, into an electron positron pair. The positron joins with the positronium atom, and we have the proton.

Solid as a rock. The proton is going nowhere !

It seems we are entering photoproduction territory here. Photoproduction, *bremsstrahlung*, it's all the same thing. Matter shedding energy trying to achieve a state of repose.

(could this be our metaphysics? I enjoy reposing, as well ... so do my dogs !)

As we clearly foresaw over twenty years ago, photoproduction is wave of the future!   :)

Photoproduction experiments and neutrino experiment; it seems they could be done together.

The Dirac equation:

The universal Dirac equation is

$$H \psi = \alpha \cdot p \, \psi \qquad (7)$$

$$i\hbar \, \partial \psi / \partial t = -i\hbar \, \alpha \cdot \nabla \psi \qquad (8)$$

and is satisfied by, or solved with, the universal wave function

$$\psi = \exp(i(p \cdot x - (m-m_0)c^2 t)/\hbar) \qquad (9)$$

$$mc^2 \psi = \alpha \cdot p \, \psi + m_e c^2 \psi \qquad (10)$$

The universal model is *the* gauge theory anticipated by the standard model !

Rest mass is the global charge; and the relativistic mass is the local, or variable charge.

We salute the constructors of the standard model for keeping this necessary,
and "beautiful", idea alive.   yay!

Unfortunately, there were a few misinterpretations on the "meaning" of the mathematics, and then, of course, everyone took the idea *way too far*, as people *always* do.

The result: Massive and charged propagators, eight bi-colored gluons, a bunch of dodgy math ... *et voila* !; you're a television spokesmodel, a prophet, and a futurist !

sweet

But, back to physics.

One free particle solution of the Dirac equation for an electron (positive helicity) is

$$\psi = \begin{vmatrix} 1 \\ 0 \\ \sigma \cdot p/mc^2 \begin{vmatrix} 1 \\ 0 \end{vmatrix} \end{vmatrix} \exp(i(p \cdot x - (m-m_0)c^2 t)/\hbar) \qquad (11)$$

Please forgive the slightly wonky notation. This is the 'spin up' solution.

The solution for the positron (negative helicity) is then

$$\psi = -\sigma \cdot p/mc^2 \begin{vmatrix} 1 \\ 0 \\ 1 \\ 0 \end{vmatrix} \exp(i(p \cdot x - (m-m_0)c^2 t)/\hbar) \qquad (12)$$

also spin up!

There are spin down solutions too, although the helicity of the electron and the positron remain the same.

We no longer need to employ the factor γ^5 to project out left handed and right handed solutions when we are studying leptonic interactions.

Thus, there are no axial vectors in the universal model.

To include the electromagnetic interaction for an electron, we write

$$i\hbar \, \partial\psi/\partial t = -i\hbar \, \alpha \cdot \nabla \psi + (e/m_0 c^2) A^\mu \, i\hbar \, \partial\psi/\partial t \qquad (13)$$

$$i\hbar \, \partial\psi/\partial t \, (1 - A^\mu e/m_0 c^2) = -i\hbar \, \alpha \cdot \nabla \psi \qquad (14)$$

Assume A^μ is due to a second electron

$$A^\mu = -im_2(e/m_0 c^2) \, 1/(v_f - v_i)^2 \, (\phi_f^*(\partial_\mu \phi_i) - (\partial_\mu \phi_f^*)\phi_i) \qquad (15)$$

We have boldly replaced q^2 with ΔV^2

Are we just making junk up as we go along ?

Sometimes !

Close enough for government work !

In equation (15), the 'propagator' term is now in terms of the covariant four velocities of one of the interacting particles. The integral is performed over the variable v_2, from v_i to v_f.

This integral should not blow up! There is no need to perform complex integrals to avoid the "mass pole".

The factor A_μ represents the wave function of the virtual photon; $A_\mu \equiv \gamma$.

$$\gamma = \int\int\int\int\int \exp(ipx - Et)/\hbar \tag{16}$$

where

$$E = m_2 c^2 - m_1 c^2 \tag{17}$$

In this description, there are no fields, no photon "exchange", and the entire interaction is described in terms of the three well behaved wave functions of the three particles involved.

the end

?

On angular acceleration:

    In the grand scheme of things, from the point of view of an universal, inertial reference frame, situated in a galaxy far, far, away, there is no need, or room, for the concepts of linear momentum and rectilinear acceleration. Linear motion is relative.

    Angular acceleration is 'true' acceleration and results in the emission of tangential, retarding radiation, guaranteeing the conservation of energy and momentum.

    In the universal reference frame, the radiating particle and the emitted photon will have equal and opposite angular momenta.

    The world is bathed in photons. From human made electromagnetic radiation, to cosmic rays, to gravitational bremsstrahlung, we are awash in light.

    Light is how we see and communicate!   Light visible and otherwise.

    In physics, we speak of real and virtual photons. There are free photons, bound photons, and binding photons. We are leptons and radiation.

    We are stardust, billion year old carbon, we are golden, caught in the devil's bargain.
                                                             -- Joni Mitchell

                                                                                                        :)

                              Step into the light !

                    We've got to get ourselves back to the garden !

                                    True, dat.

                What's so funny 'bout peace, love, and understanding ? -- Nick Lowe

                nothing   (not really a *non sequitur*)   #introvertsdostuff   !

Conclusion:

People are not rational.   I am not rational !

I have many irrational thoughts a day. Sometimes, six before breakfast !

Let us say rationality consists in identifying, collecting, and ridding ourselves of irrational thoughts; a *constant* housecleaning, a *victorious* circle of logic, reason, and reflection.

A task we can only complete together.

A tricky task, as it seems there is no common, human nature to bind and aid us.

Imagine what it must be like to inhabit the fevered, syphilitic, dream of the celebrity physicist. A place where logic, reason, and good taste go to languish and die.

Impossible !

                                     LOL   :)

Restoration:

Let us return logic, reason, and rationality to the throne; these are the crown jewels of being. Let common sense, again, guide us in our human, and scientific, endeavors.

Are they not one and the same ? Is there really anything else ?

                                                         . . .   only succumbing to the mob

Resources:

Quantum Field Theory   (saved our electron magnetic moment bacon !)
Claude Itzykson, Jean-Bernard Zuber

Atomic and Quantum Physics
H. Haken, H.C. Wolf

Modern Elementary Particle Physics
Gordon Kane

Classical Dynamics of Particles and Systems
Jerry B. Marion

Foundations of Electromagnetic Theory
John R. Reitz, Frederick J. Milford, Robert W. Christy

Quantum Physics (awesome)
Rolf G. Winter

Gauge Theories in Particle Physics
I. J. R. Aitchison and A. J. G. Hey

Quarks and Leptons: An Introductory Course in Modern Particle Physics
Francis Halzen, Alan D. Martin   (most excellent !)

Quantum Field Theory
F. Mandl, G. Shaw

Theoretical Mechanics of Particles and Continua
Alexander L. Fetter, John Dirk Walecka

The Theory of Spinors
Elie Cartan

Elementary Modern Physics   (Best Book Ever!)
Richard T. Weidner, Robert L. Sells

Quantum Mechanics
Claude Cohen-Tannoudji, Bernard Diu, Franck Laloe

Books by Greg Feild: The SInister Universe Series

the pentateuch

1. "A quantum mechanical theory of gravitational interactions"
   CreateSpace Independent Publishing, 8/29/2016

2. "Observations on the quantum mechanical nature of gravity"
   CreateSpace Independent Publishing, 10/8/2016

3. "On gravitation and electric charge"
   CreateSpace Independent Publishing, 10/29/2016

4. "On spin, mass, and charge"
   CreateSpace Independent Publishing, 11/29/2016

5. "On angular momentum, acceleration, and absolute motion"
   CreateSpace Independent Publishing, 1/1/2017

the exegeses

6. "The Sinister Universe"
   CreateSpace Independent Publishing, 3/1/2017

7. "On Parity and Isospin"
   CreateSpace Independent Publishing, 4/11/2017

8. "Reflections on the Sinister Universe"
   CreateSpace Independent Publishing, 5/12/2017

the hermeneutics

9. "On Current Physics"
   CreateSpace Independent Publishing, 6/11/2017

10. "A Critical Examination of Classical and Quantum Mechanical Waves"
    CreateSpace Independent Publishing, 6/18/2017

the gospels :)

11. "On wave particle duality and the quantum of action"
    CreateSpace Independent Publishing, 7/6/2017

12. "On matter, mass, and motion"
    CreateSpace Independent Publishing, 9/14/2017

13. "On action and reaction"
    CreateSpace Independent Publishing, 9/24/2017

14. "A quantum mechanical theory of everything"
    CreateSpace Independent Publishing, 11/5/2017

the expositions

15. "On Interaction"
    CreateSpace Independent Publishing, 4/21/2018

16. "On Rotation"
    CreateSpace Independent Publishing 8/19/2018

the matchbook summary

17. "Revenge of the Sinister Universe: The Reality of Everything'
    CreateSpace Independent Publishing, 9/4/2018

the diatribe

18. "On Math, Physics, and Metaphysics"
    CreateSpace Independent Publishing, 10/1/2018

the compilations

"The Universal Model of Our Sinister Universe: The First Ten Books"
CreateSpace Independent Publishing, 7/2/2017

"The Canons of the Sinister Universe:
The Last Four Books on the Universal Model of Our World"
CreateSpace Independent Publishing, 11/5/2017

"The Return of the Sinister Universe: The Immaculate Collection"
CreateSpace Independent Publishing, 9/4/2018

Notes:

hell is other people

what can you do ?

try to enjoy !

there is only <u>now</u>

:)